ONE.

ONE.

Reclaiming Humanity in the Age of Superintelligence

Jawna Standish & Joshua Badshah

Published by Noora Labs, an imprint of ONE.
Joshua Badshah, Author photo (c) Pingping Xiao and VibeShack Studio
Jawna Standish, Author photo (c) Kyran Daniel and Noora Labs

Paperback ISBN: 979-8-9932499-1-9
Hardcover ISBN: 979-8-9932499-0-2
Digital ISBN: 979-8-9932499-3-3

From Jawna:

To Ryanne and Elias, whose presence and generosity remind me that love is our greatest technology. And to the stardust that connects us all, the quiet proof that we truly are all ONE.

From Joshua:

To Catherine (my piano teacher) and Graham. Without my music, none of this would have unfolded like it did. My deep gratitude to Dr. Kirchner and Mummy & Daddy.

Read This First

Thank you for supporting and reading our book.
We'd love to connect with you on social media:

Scan our QR Code Here:

https://qrkit.co/kKA2Tf

ONE.

RECLAIMING HUMANITY
IN THE AGE OF
SUPERINTELLIGENCE

JAWNA
STANDISH   &   JOSHUA
BADSHAH

Table of Contents

Introduction 1

1. Consciousness 3

2. Programming 25

3. Self 37

4. Wisdom 49

5. Alignment 61

6. Love 71

7. Community 83

8. Sound 97

9. Within 105

10. The Polymath 115

11. Essence 123

Conclusion 131

Introduction

From Cosmos to Consciousness to Control

Over five thousand years ago, the ancient Indian Vedas were already describing something modern science is still struggling to name, a field of awareness beyond matter. Long before the word *consciousness* existed in the West, the Vedas spoke of *cit*, pure consciousness, the ground of being, as the essence behind thought and perception.

Two thousand years ago, the Greek philosopher Plato wrestled with the same mystery: *What is consciousness?*

He argued that there was a gap between the physical world and the world of the mind, between what we can measure and what we can feel.

In many ways, we're still right where they left us.
Today, neuroscientists echo his dilemma: open up a brain and you find molecules and firing neurons, but never a thought, never love, never the color red or the sound of laughter. No scientist has ever discovered an idea tucked between synapses.

This is what scientists and philosophers call the *hard problem of consciousness.*

And here's why it matters now: we're entering an age of superintelligence. An age where machines don't just run our newsfeeds but are being woven into our bodies, our biology, and our choices.

The danger isn't that AI will become "too powerful."

The danger is that we'll stop questioning it. That we'll accept its outputs as truth.

That we'll forget the mystery of our own minds, and with it, our agency, creativity, and humanity.

If we can't even explain how consciousness emerges, why would we blindly assume a machine, AI, and eventually superintelligence are capable of telling us who we are or what's real?

This is not a book about the mechanics or technology behind AI and superintelligence.

Rather, it's about stepping into one of humanity's ancient conversations, the mystery of consciousness, and why understanding it now is critical to how you live, love, and lead in this new era of superintelligence.

Because it's not coming soon. It's already here, altering our brains and biological responses at a cellular level.

This is not a fear-mongering takedown of AI or superintelligence; after all, we are scientists and technologists who believe in technology and the human advancements that have come about as a result. However, we are also artists, creatives, and seekers.

As we each journeyed to explore our own existence, we were reminded of how much technology shapes everything we do today.

We were also reminded how much ancient wisdom has to offer. And how much of it we have forgotten. That's why it has stood the test of time.

This is not a science or textbook. Instead, it's a story: part science, part art, part call to stay profoundly human.

We hope you'll return to it whenever the future starts to feel too machine-made.

As we began, we explored some of the unanswered questions in science today. Though we tend to assume technology can solve everything, it can't.

Not yet.

The Cosmos Reminds Us Who We Are

To ground you in consciousness, we had to look at the universe's composition.

Seventy percent of the universe is made of something scientists call *dark energy*: a force pushing galaxies apart faster than the speed of light.

We don't know what it is. Einstein once dismissed it as his "biggest mistake."

It wasn't.

Another twenty-seven percent is *dark matter*: invisible, yet responsible for nearly all the gravity that holds galaxies together. Without it, Earth itself would drift into the void.

That leaves just three percent, the so-called *atomic universe*. And of that? 99.999 percent is dust.

This means that everything we can touch, see, and measure, including our bodies, cities, mountains, and even the stars, makes up less than a fraction of a fraction of reality.

And even atoms aren't solid. They dissolve into waves the moment no one looks.

Ask a physicist where those waves come from, and they'll point to "quantum fields" or "Hilbert space." But the truth is simpler and stranger: the universe is born from imagination.

So if you ask me, "What is the universe made of?" The most honest answer is nothing.

And yet, here it is. Here we are.

Which circles us back to Plato.

If the universe is made of nothing, why does it look like something?

Why does it look like *this*, like you, like me, like a child's laughter, or the beauty of a night sky?

The Spark

That question became the heartbeat of this book.

Because beneath the algorithms, beyond the science, past the labels of scientist or artist, seeker or builder, we both recognized this: humanity is the spark. The stardust. The consciousness. The mystery. The part no machine can replicate.

And once you see that, everything changes.

Because this book isn't about technology. It's about you, me, and everyone.

We really all are individuated pieces of ONE thing, the same stardust.

It's about the way you move through the world when you remember you're not just reacting to inputs or living on autopilot. You're choosing, creating, imagining.

When you see the mystery of consciousness clearly, the fact that no telescope has ever revealed the essence of love, no algorithm has ever generated awe, you stop worrying about whether machines will replace humans. You understand they can't.

They don't carry what you carry.
They don't hold soul.
They don't generate wonder.
They don't wake up aching for connection.

That realization is the awakening this book invites:

- To step into this age of superintelligence with your eyes wide open.
- To embrace AI as a tool but never mistake it for the source of meaning.
- To reclaim the uniquely human capacities of intuition, awe, creativity, and love that so many are just blindly and unknowingly outsourcing to AI without so much as a thought.

Once you see this, you no longer fear AI as competition.
You see it for what it is: fire.

Powerful, yes.

World-changing, yes.

But still just fire, waiting for human hands, human hearts, and human consciousness to decide what it will create or destroy.

And the closer we got to that realization, the more it began to ask something of us.

Choices that felt less like logic and more like risk. Less like certainty, more like faith.

Because reclaiming humanity in the age of superintelligence isn't about tech, ethics, religion, politics, or which generation you were born into.

It's none of those things.

It's about remembering we are not passengers.

We are not data points to be optimized.
We are not background noise in the great algorithm.

We are the question.
We are the mystery.
We are the love.
We are the consciousness that machines will never replicate.

And once you see that, you stop living in fear and on autopilot.
You stop creating an incoherence between your brain and your heart.

Your body then stops producing the hormones and chemicals related to fear and disconnect.

You stop outsourcing your free will and creative choices.

You stop confusing convenience with freedom.

You can begin to create from a new place of joy, love, freedom, and abundance.

Because once you start asking what's real, you can't stop. And what we discovered next would test everything we thought we knew — about algorithms, about intuition, about ourselves.

Let's begin.

With love,
Jawna & Joshua

"

"As your heart desires… make us a city for the ages." — Athena

1

Consciousness

A Message That Changed Everything

I, (Jawna) still remember the moment I received the DM on LinkedIn.

At first, I didn't even think it was from a real person.

It's social media, after all: half bots, half noise, and only sometimes real connection.

The irony is that I hadn't used LinkedIn in a while, but decided to take an open-minded approach to leveraging it and committed to consistently posting several days a week for nine months. I could see it was making a difference in my business and with the people I was meeting. My intentional visibility definitely made a difference in the opportunities that I was attracting.

Something about this DM stopped me.

The man's photo looked like the same man I'd noticed at the longevity sciences conference at Stanford University School of Medicine a few weeks earlier, the one in the back of the room who kept catching my eye during the coffee break, as though the universe was insisting our paths cross.

As I was walking over towards him to say hello, one of the speakers came to introduce themselves to me, and as life would sometimes have it, I was redirected. So we never found each other again that day.

So when this message appeared in my inbox, I thought, *Maybe it's him. Maybe synchronicity is finding a way.*

The truth is, just a few months earlier, I might have ignored the DM altogether. And perhaps the man sending it may not have done so at all had he also not been on his own journey.

I was freshly back from a spiritual and scientific journey that cracked me wide open.

I had just spent a week at one of Dr. Joe Dispenza's advanced meditation retreats, as a neuroscientist and as a guest, where participants were wired to EEGs, heart monitors, and even selected for blood panels before and after meditation to track the biological and chemical changes happening in the body during meditation.

I watched brainwaves and heartbeats sync in real time. I saw metabolites in the blood shift after deep meditation. Proof that magic doesn't just live in metaphor—it shows up in molecules. It's not *woo-woo*. It's neuroscience. It's physics. It's chemistry. It's biology. It's consciousness at its finest.

That retreat left me with a conviction: consciousness isn't abstract. Not at all.

It isn't just a word philosophers, neuroscientists, and physicists toss around. It's alive, shaping our choices and paths in ways we barely notice until we look back. Without consciousness, there is no sustainable success in any area of life, including love, health, relationships, and one that people often overlook: your business. It's that profound. Yet it's not discussed enough.

So when the DM appeared that day on my LinkedIn, something in me whispered: *Reply, Jawna.*

A Life Between Science and Humanity

Just months before that moment, I was living on a sailboat in Ikaria, Greece, one of the world's "Blue Zones." We humans have always tried to map wonder.

Sometimes it's stars. Sometimes it's seas. And sometimes it's the rare places where life lasts far longer than we expect.

What's a Blue Zone?

The term "Blue Zones" began quietly in 2004, when researchers studying longevity in Sardinia, Italy, circled clusters of centenarians on a map with a blue pen. That simple mark gave a name to something profound: regions where people live not just long, but *well*.

National Geographic explorer Dan Buettner later helped the idea travel the world, identifying five remarkable hotspots: Sardinia,

Okinawa, Nicoya, Ikaria (where I lived and researched), and Loma Linda.

But beyond the books and Netflix series, the heart of Blue Zones is simple and deeply human.

They're not about gadgets or hacks. They're about ordinary rhythms that add up to extraordinary lives: natural movement, nourishing food, purpose, close relationships, time to belong.

In an age racing toward AI and superintelligence, these places remind us that longevity isn't just *time*; it's aliveness. And staying "human" might be our greatest technology.

In Ikaria, Greece, I lived and researched from my sailboat, my floating lab, in rhythm with this truth.

My days began with morning swims, afternoons with pottery, painting, making perfume, and evenings stretched long into the night with family, friends, and neighbors—meals, music, sometimes dancing with villagers.

We were a multigenerational amalgamation: from kids to those well into their eighties, nineties, and, of course, centenarians. Our very lives were evidence of resilience, rhythm, and love.

Meanwhile, on the other side of the world, Joshua, the man who sent me the DM on LinkedIn, was walking his own pilgrimage.

By day, he's a Stanford University School of Medicine scientist, longevity researcher, and bioinformatician in stem cell research and the reversal of aging. By night, a poet, musician, and filmmaker.

He had just returned from Thailand, where he studied scriptures and the silent mystery of consciousness beyond words.

At the time, we still didn't know each other.

Not yet.

And that is the paradox of consciousness: it is both deeply personal and universally shared.

We both joked that LinkedIn was the "dead internet." Clearly, we were wrong about LinkedIn.

We've talked about it many times since, laughing that I probably wouldn't respond to his DM today, given where our lives are now, and that he wouldn't be sending DMs either.

Was it synchronicity? Timing? The LinkedIn algorithm? The universe winking?

Maybe all of the above.

We wanted to explore this, and that's how our collaboration projects, including this book, began.

We witnessed from our own lived experiences how each of our lives had improved and expanded when we increased our consciousness. And we wanted to help others increase theirs, too.

02:19

.ıl LTE 92

← Joshua Badshah ··· ◻︎ ☆

Joshua Badshah · 1st

Life Science Research Professional at Stanford University School of Medicine. Co-Founder Daiyan Solutions.

JAN 23

Joshua Badshah · 8:09pm
Possible collaboration
Hey Jawna,

I am a research scientist at Stanford University, in the field of ageing. I would love to connect with you, hear from you, and see if we could maybe work on something.

I have lived in 5 countries, and just moved to the Bay 2 years ago from Aus. I have extensive experience with start ups. And want to change the world of science.

⌷ Write a message... 🎤

02:19 ✈   .ıl LTE 91

←   Joshua Badshah   · · ·   📷   ☆

I have lived in 5 countries, and just moved to the Bay 2 years ago from Aus. I have extensive experience with start ups. And want to change the world of science.

Would love to hear back from you,

Joshua

Jawna Standish · 8:14 pm
Hi Joshua, Thanks so much for your nice note! Happy to be connected & in each others' networks. I'd love to learn more about your work & yes I'm always open to collaborating! I was just at Stanford School of Medicine last week for the Longevity Symposium hosted by Dr. Ronjon Nag. By any chance were you there, too?

Jawna Standish · 8:16 pm
I'm not currently in Palo Alto, so if

📎   Write a message...   🎤

The Myth of Fire

If there was one moment that ignited not just our survival, but our consciousness as humans, it was the discovery of fire.

Greek mythology tells of Prometheus, the Titan who defied the gods by stealing fire from Olympus and gifting it to humanity.

Zeus punished him brutally, but the myth endured for millennia because it was true on some deeper level.

Fire was rebellion.

Fire was possibility.

Fire was the technology that made us human.

Before fire, our ancestors lived like most mammals: reactive, impulsive, always on the edge of survival. Eat or be eaten. Stay warm. Procreate. Stay alive.

Then came fire—1.5 million years ago.

It gave us light, warmth, and protection from predators.

It was the catalyst that transformed raw food into digestible calories, unlocking nutrients our ancestors had never accessed.

Less energy was required for digestion, freeing up surplus calories. If you've ever heard that the gut and the brain are connected, that's correct; they are indeed. We'll get into more of that in the coming pages.

What's most important to understand about fire and human evolution is that fire cooked our food, and evolution invested those calories in the formation of a new part of the brain.

That surplus grew into what's called your prefrontal cortex: the crowning jewel of human evolution.

The PFC gave us impulse control, imagination, empathy, storytelling, and planning.

It allowed us to delay gratification, envision the future, and organize societies.

Fire didn't just warm our bodies. It lit up our minds, quite literally and figuratively.

And perhaps the greatest gift fire gave us?

Time.

Does any of this sound remotely familiar to what AI may be doing for humanity today? Stick with us. It gets even better.

Let's go back to the cave days. For the first time, humans gathered around a fire at night. Not just surviving, but reflecting, dreaming, telling stories.

Storytelling became our superpower.

It allowed us to speak, to pass knowledge across generations, to create myths and meaning, to weave values that held tribes together.

Fire made us human.

And today?

AI is our modern fire.

It's reshaping our brains, our societies, our consciousness, just as fire did. But at a speed evolution never prepared us for.

The question is not *whether* it will change us.

It's whether we'll remain awake to what that change means and if we'll embrace it consciously.

Your Amazing Brain: An Evolutionary Stack

Your brain is not one machine. It's an evolutionary stack, meaning it's built layer by layer over millions of years.

Brainstem (Reptilian Brain)

- Approximately three hundred million years old.
- Governs survival: breathing, heartbeat, digestion, sleep, arousal.
- Instinctual, automatic, always on.
- From a consciousness perspective, this is where *interoception* begins: the awareness of your inner world. Hunger, thirst, heartbeat, gut feelings, they all start here.

Limbic System (Mammalian Brain)

- Added with mammals.
- Governs bonding, memory, emotion, motivation.

- Structures: amygdala (fear/threat detection), hippocampus (memory), hypothalamus (hormones).
- This is why you cry at movies, miss someone you love, or feel nostalgia when you smell your grandmother's cooking.

Neocortex / Prefrontal Cortex (Human Brain)

- The newest layer, our evolutionary edge.
- Governs planning, empathy, abstract thought, imagination, storytelling.
- Expanded after fire, when cooked food freed energy for brain growth.
- This is the layer that made civilizations, art, and algorithms possible.

Here's the irony: these layers still operate together today.

Under stress, your ancient brainstem and limbic system can hijack the logical PFC in milliseconds. That's why you can "know" something rationally but still act out of fear or impulse.

And in the age of AI, where algorithms are engineered to hijack your limbic system with notifications, dopamine hits, and nudges, understanding this stack isn't optional.

If you don't know how your brain works, someone (or something) else will know it better than you.

Limbic System

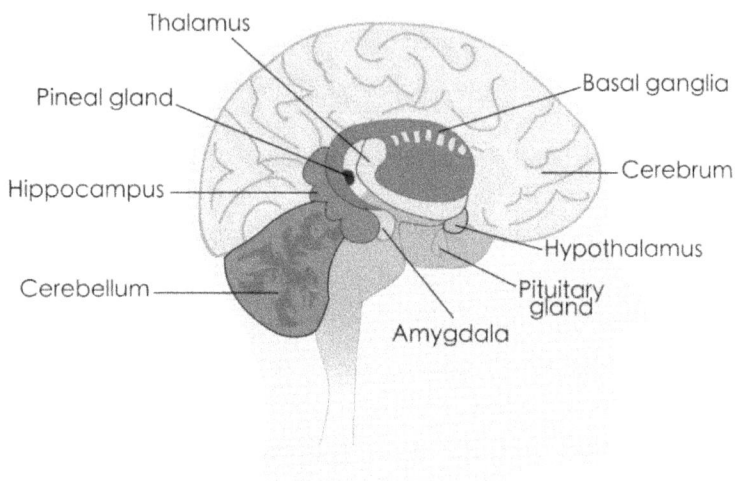

Interoception: The Forgotten Super Sense

Most of us were taught we have five senses: sight, sound, taste, touch, and smell.

But modern neuroscience shows we actually have at least thirty-four senses. Most aren't ready to hear that, as it goes against everything you were always taught. But science and times are changing. The reality is, we don't know what we don't know.

Interoception covers many of these additional senses happening in the body: sensing blood sugar, detecting heart rate, feeling muscle tension, and noticing internal temperature shifts.

You don't "see" your blood pressure, but you feel stress. You don't "hear" low blood sugar, but you sense the crash.

Interoception is the body whispering its truth before your conscious brain catches up.

It's the butterflies in your stomach when you're nervous.
The lump in your throat when you hold back tears.
The gut feeling that warns you before your mind has data.

And here's why it matters: interoception is the bridge between biology and meaning.

Without it, you wouldn't know joy, longing, or love. Machines don't have that.

Algorithms can't replicate it.

This (your ability to feel your inner world) is part of what makes you irreplaceable in the age of superintelligence.

Consciousness Through Time: A Timeline

We humans have always tried to map wonder and the invisible.

Sometimes it's stars. Sometimes it's seas. And sometimes it's the invisible life inside us, the mystery of being aware.

1500–500 BCE: The Vedas & Upanishads (Ancient India)
Explored *Ātman*, the self, and its connection to *Brahman*, universal consciousness.
Taught that awareness is fundamental, and meditation is a way to know it.

---

400 BCE: Plato (Greece)
Argued that mind and soul couldn't be reduced to matter.

1600s: René Descartes (France)
"Cogito, ergo sum." *I think, therefore I am.* Thinking as proof of existence.

1700s: Hume & Kant (Scotland & Prussia)
Hume doubted a permanent self; Kant claimed the mind shapes reality.

1800s: William James (United States)
Described consciousness as a "stream." Helped turn psychology into a science.

Early 1900s: Carl Jung (Switzerland)
Explored the collective unconscious, archetypes, and myth across cultures.

Mid-1900s: John von Neumann (Hungary/US)
Linked quantum theory, computing, and game theory; warned of a coming "singularity."

2005: Ray Kurzweil (United States)

Futurist and author of *The Singularity Is Near*, predicted that by 2045, machine intelligence could surpass human intelligence.

For millennia, we've wrestled with the mystery of awareness, from the sages of India to the philosophers of Greece to the scientists and technologists shaping today's world.

It's this same question that pulled us to write this book.

Because as AI races toward superintelligence, the conversation is no longer just about *what consciousness is.*

It's how we keep humanity alive, rich, and fully our own, and evolve it in an age when machines will outthink us.

Consciousness as Energy

Here's something most people never stop to think about: consciousness has a frequency.

Dr. David Hawkins' *Map of Consciousness* organizes human states of being into measurable levels of energy.

At the bottom: shame, guilt, fear, anger, which are contracting, survival states.
In the middle: courage, neutrality, willingness, acceptance.
At the top: love, joy, peace, enlightenment, which are expanding, creative states.

Your nervous system knows these frequencies instantly. You've felt it:

- The heaviness around someone stuck in anger.
- The calm with someone grounded in compassion.
- The electric buzz around people lit up with joy or purpose.

That's not *woo-woo*—it's neurobiology.

Hawkins argued humanity's global average sits around 200 to 250. That means most people live between courage and neutrality. Not bad, but far from our full potential as humans.

And here's the good news: consciousness is fluid. You can shift it. You can elevate it at any time.

Level	Energetic Vibration	Emotional State	World View
Enlightenment	700–1,000	Ineffable	The highest state of pure consciousness and spiritual realization.
Peace	600	Bliss	Unity and transcendence.
Joy	540	Serenity	Compassion, love, and a feeling of unconditional happiness.
Love	500	Reverence	Unconditional love, compassion, and forgiveness.
Reason	400	Understanding	Objective and rational thinking, which moves beyond emotional responses.
Acceptance	350	Forgiveness	Taking responsibility for one's life and transcending emotionalism.
Willingness	310	Optimism	Actively participating in life and being proactive.

Neutrality	250	Trust	Flexible, detached, and non-judgmental.
Courage	200	Affirmation	This is the critical threshold, or fulcrum, where the energy shifts from destructive to constructive power.
Pride	175	Scorn	Feeling superior, but this state is fragile and defensive.
Anger	150	Hate	Frustration arising from desires not being met.
Desire	125	Craving	A state of needing and longing; can be a driver for addiction.
Fear	100	Anxiety	A constant sense of worry, anxiety, and insecurity.
Grief	75	Regret	Sadness, loss, and mourning.
Apathy	50	Despair	Indifference, hopelessness, and low energy.
Guilt	30	Blame	Feelings of remorse, self-blame, and condemnation.
Shame	20	Humiliation	The lowest level, characterized by despair and humiliation.

Chapter Summary

- Consciousness is humanity's spark. Machines can't replicate it.

- Fire grew our PFC, gave us time, and birthed storytelling. AI is our new fire.

- Your brain is an evolutionary stack: reptilian, mammalian, human. Stress can pull you back into survival mode in milliseconds.

- Interoception, your hidden "thirty-four senses," is the bridge between body and awareness.

- Consciousness has a frequency. Most people live around 200 to 250 (from courage to neutrality). But love, joy, and creativity are available if we rise.

- From the Vedas to Kurzweil, humanity has been wrestling with consciousness for millennia. You're stepping into that lineage now.

Field Notes: Consciousness Practices for the Week

1. **Notice Your Brain Stack**
   - o Ask: Is this my brainstem (survival), my limbic system (emotion), or my PFC (logic/imagination) running the show right now? Naming it gives you choice.

2. **Interoception Check-In**
   - o Pause once today. Ask: What is my body telling me? Notice heartbeat, tension, sweaty palms, gut, breath. Don't judge, just listen.

3. **Track a Synchronicity**
   - o Write down or take a voice note during one moment of synchronicity this week. Don't analyze. Just notice.

4. **Frequency Shift**
   - o If you catch yourself in fear or anger, pause. Ask: What small step could shift me toward courage, acceptance, or love right now?

Closing

Consciousness is not just neurons firing. It's the lived experience of being human and being aware of yourself and how you move through and impact others in this vast, big universe. It's what makes you who you are.

Fire once expanded our minds and rewired our evolution.

AI is our new fire.

The only question is: will we remain awake enough to use it consciously, guided by the spark no machine can replicate: awareness, awe, creativity, and love?

"

"Know thyself."
– Delphic Maxim

2

Programming

The Myth of the Labyrinth

In Greek mythology, King Minos built a labyrinth so intricate that no one who entered could escape. At its center lived the Minotaur, a beast that demanded sacrifice.

The labyrinth wasn't just stone walls; it was a system of control. Minos knew that as long as the people believed the maze was inescapable, they would submit to his authority.

It took Theseus, with the help of Ariadne's thread, to change the story. By holding onto that single thread of truth, he found his way out.

Fast forward to today: our digital lives are a modern labyrinth. Except the walls aren't stone—they're algorithms. And the Minotaur isn't a beast; it's centralized systems: corporations, governments, and financial institutions that hold the keys.

The question is: *Who holds the thread?*

Centralization: The Default Program

In the world we live in today, most of our technology and organizational structures are centralized. What does that mean, and how is it related to consciousness, you may be asking?

As it turns out, it has everything to do with us: how we think, act, and feel. Because a few corporations, governments, leaders, and entities own the servers, the code, the final decisions, and most importantly, your data.

Let's use an example most of us can relate to in this modern day. Think about your wearable device, or simply your smartphone that you carry everywhere, to track your wellness and fitness—your Oura ring, WHOOP band, Apple Watch, or Garmin. They have the ability to track your heart rate, your sleep, your mood, your menstrual cycle, even when you're ovulating.

That data should belong to you. It was generated by you.

And yet…

- You don't own it.
- You don't profit from it.
- You have no control over how it's used.

That information flows into big corporate silos, fueling a multi-trillion-dollar industry called *longevity*. Your body, at a cellular level, is literally building someone else's empire. Stop and think for a moment.

This is the modern labyrinth: your most intimate truths, your habits, your inner code, your biology, locked away, repackaged, and sold back to you as a subscription service you pay for life with little to no say in how any of it comes about.

And here's the "aha" moment most people miss: you've been programmed to believe this is normal and that it's good for you. After all, you want to be healthy, right?

You've been trained to trade convenience for self-sovereignty.

To hand over ownership of your data, in this case, in exchange for access. To confuse "free" with freedom.

But what if that's not the only way?

Decentralization: A New Thread

That's why new technologies and movements are emerging. Blockchain. Crypto. Distributed networks. Balaji Srinivasan's *The Network State*.

All of these are examples of social movements, designed by humans. Instead of trusting a single authority, blockchain relies on a distributed network of validation. No one owns the ledger; everyone verifies it.

That means you could hold the keys to your own data. You decide what's shared, with whom, and when. You could even monetize it if you choose, because it's yours.

This isn't just about cryptocurrency or NFTs. At its core, decentralization is about self-sovereignty, opportunity, and trust. It's about reclaiming your inner code—your biology, beliefs, and choices—from the very systems that have been designed to profit off your unconsciousness.

It's Ariadne's thread in the modern labyrinth.

Within Blockchain, Here Are How Smart Contracts Work (In Plain Language)

To really understand decentralized blockchain technology, you need to understand smart contracts.

Think of them as "digital vending machines."

When you put money into a vending machine and press a button, you don't need to trust a cashier. The machine automatically delivers the snack if the conditions are met.

Smart contracts work the same way. They're pieces of code on the blockchain that automatically execute agreements when the conditions are met.

- You send tokens to pay for the goods (the input).
- The smart contract digitally checks the rules for that transaction.
- If the conditions are true, it executes the outcome automatically.

No banker. No lawyer. No corporate intermediary.

Now imagine this with your health data.

You could program a smart contract that only releases your sleep data to a research lab if they pay you for it, and only for the use and duration you choose. You could prevent it from being used for military research and development, for example. When the conditions expire, access shuts off. No fine print. No hidden clauses. Automatically.

Smart contracts wouldn't just eliminate middlemen. They would return control to you.

Humans as Storytelling Beings

Up until forty thousand years ago, there were eight different species, including us, Homo sapiens.

We all have a language that we understand amongst our own species.

They all center around three fundamentals that are tied to our survival:

1. Food calls
2. Mating calls
3. Danger calls

Whales, cats, dogs, and birds all have their own languages.

But only one species discovered something different. Only we (Homo sapiens) developed the ability to tell stories, thanks to our PFC. To radiate them. To perpetuate them.

Why? Because fire grew our PFC: the part of our brain that allows us to imagine, to delay gratification, to strategize, and to tell stories.

And stories changed everything for humanity.

- Money is a story. A collective agreement about value exchange.
- Nation-states are stories. Borders drawn in ink and blood that only exist because we all agree they do. Think about it: why do you really need a passport?
- The Bible, the Koran, the Vedas, the Tripitaka—all stories. Sacred ones. They have moved billions of people across centuries, inspiring devotion, sacrifice, and transformation.

Stories are not illusions.

They are the most powerful human technology. They create reality at scale.

And here's the beautiful thing: humans are still telling new stories. About money (think blockchain). About nations (think Balaji Srinivasan's *The Network State*). About new experimental societies like Prospera. About micro-communities forming online, gathering around new, shared values, and for those interested in new forms of governance in digitally native states and crypto-powered societies.

Despite what society and traditional media may say, we are not trapped in old myths. We have the ability to author new stories.

The question is, will we?

AI as Fire

This is where the conversation shifts. Ask ten people and you'll get ten different answers.

"What do you think about AI?"

AI is not the enemy. It's fire.

Like fire, AI can destroy—or create.
Like fire, it can save lives—or burn down civilizations.
Like fire, it gives us a choice.

AI can free your headspace.

It can save you hours.

It can help you learn faster, write more, and design with more creativity.

It can give you more time to focus on your zone of genius—the part only you can do.

AI has the potential to bring abundance in ways most of us have never imagined:

- Education: Personalized learning at scale.
- Health: Early diagnostics, mental health support, and prevention.
- Financial freedom: Access to tools and opportunities once reserved for elites.

But here's the catch: AI is not intelligent. You are.

AI is fire waiting for human hands, human hearts, human consciousness. It amplifies what you already carry. If you forget that, you hand over the match to someone else.

AI Is Fire

- Fire can destroy. It can burn down forests, homes, and civilizations.
- Fire can create. It cooks food, forges steel, and warms families.
- Fire gave us time. It freed calories to grow our brains, tell stories, and imagine futures.

AI is the same

- It can enslave your attention.
- Or it can free your mind.
- It can trap you in someone else's story.
- Or it can help you rewrite your own.

The difference? Conscious use.

The thread of control has always been in human hands.

The question is not whether AI will change us.
The question is, will we remain awake to what that change means?

Predictive Consciousness Systems

This is where the conversation gets even more urgent.

Today, predictive consciousness systems are being designed to model your emotional and cognitive responses before you're aware of them.

By analyzing biometric signals and data like pupil dilation, micro-expressions, heart rate variability, menstrual cycles, peri-menopause, menopause, testosterone levels, etc., algorithms can predict with eerie accuracy how you'll respond to content, ads, even political messages. Think about that for a minute.

It's not just personalization. It's preemption.

Imagine scrolling your feed and thinking you chose to click something?

In reality, the system already predicted your click and nudged you toward that choice days earlier.

This is why consciousness matters.

If you're not aware of your own signals, someone else will be.

The Biological Revolution

Meanwhile, in the physical realm, researchers are piloting synthetic wombs—biotechnological systems designed to grow embryos outside the human body in controlled, "artificial womb" environments. You read that correctly.

Proponents argue they could reduce maternal mortality, give options to those unable to conceive, and revolutionize reproductive health.

But they also raise profound ethical questions:

- What happens to human connection and epigenetics if pregnancy is outsourced?
- What does bonding look like when a baby never hears its mother's heartbeat in utero?
- How do we redefine "human" when biology itself becomes optional?

Synthetic wombs are more than medical marvels.

They're symbols of a deeper question: will we decentralize life itself or hand it over to centralized labs and corporations?

Chapter Summary

- Centralization is today's labyrinth. Few control the walls, and your data is the sacrifice.

- Decentralization is Ariadne's thread. It returns ownership, agency, and sovereignty to you.

- Smart contracts are digital vending machines within the blockchain: code that executes agreements automatically without middlemen.

- Humans are the only species that tell stories, and those stories build civilizations, money, religions, families, and nations.

- AI is not the enemy. It is fire. It can burn or illuminate, enslave or free. The difference is conscious use.

- Predictive systems and synthetic biology force us to ask: Will we remain sovereign, or surrender our humanity?

Field Notes: Practices for the Week

Trace Your Data
Pick one app or device you use daily (like your wearable, menstrual tracking, or social feeds). Ask: Who owns this data? Who profits from it? Do I consent to that?

Notice the Stories You Live In
What story is driving your daily life: money, nation, faith, community? Write down one new story you'd like to live in.

Where Is AI Fire for You?
Ask: Where in my life is AI burning me (distraction, overload)? Where is it lighting me (support, efficiency)? How can I shift it toward creation?

Reclaim the Thread
Identify one area of your life, such as technology, health, or finance, where you feel trapped in a labyrinth. What could decentralization, or even a shift in awareness (i.e., consciousness), offer as your Ariadne's thread?

66

"The only true wisdom is in knowing you know nothing." — Socrates

3

Self

The Myth of the Oracle

Long before algorithms, long before platforms, and long before nations even existed, people still traveled long distances to ask one question: *"Who am I?"*

In ancient Greece, seekers would make pilgrimages to the Temple of Apollo at Delphi. They climbed mountains, crossed seas, and endured hardship for the chance to consult the Oracle.

Kings, soldiers, lovers, and philosophers all came, hoping for answers about their destiny.

But before they could enter the sacred temple, they passed an inscription carved into stone:

"Know thyself."

It wasn't a suggestion. It was the price of admission.

Because without knowing yourself, no prophecy mattered.

Without inner clarity, even divine wisdom was noise.

Thousands of years later, we're still seeking the same thing—to know thyself.

But instead of oracles, we open apps.

Instead of sacred inscriptions, we get slogans.

Instead of silence and reflection, we get endless notifications and confuse activity with purpose.

And yet—the ancient question still haunts us:
Who am I, really, when the world stops telling me who to be?

Self Is Not a Brand

We live in an age of identity curation. Bios polished like resumes. Photos filtered like advertisements. We construct personal brands not just for companies to notice us, but for friends, strangers, and, if we're being honest, even for ourselves.

There's nothing inherently wrong with self-expression.

Humans have always signaled identity through clothes, stories, music, or rituals.

But something has shifted. Branding has become a substitute for being.

Here's the truth most people avoid:

- The Self is not what gets likes.
- The Self is not a headline.
- The Self is not what fits neatly into 280 characters.

The Self is what remains when no one's watching.

The you when it's quiet and no one's watching.

The part of you that doesn't need applause or validation.

And if you lose contact with it, you risk building a life that looks right on the outside but feels hollow on the inside.

Many founders, executives, and medical professionals we've worked with have reached that breaking point. It's easy to get swept up in that because once you become part of a system, you must comply and conform to exist and thrive within it. It's a paradox, which is why many struggle and burn out once in these systems.

So many have spent decades climbing corporate ladders, enduring hierarchical bullying, and turning a blind eye to the unethical in the name of keeping the peace (and your job). And so many have invested over a decade in medical school and residency, only to look around and ask: *Whose life is this?* Many begin to feel like they've enslaved themselves to these institutions and have no options.

But there's always a choice.

The same story shows up in founders, creators, parents, and even seekers.

Without Self, success becomes performance, and performance without purpose eventually collapses.

Self Versus Programming

In Chapter 2, we explored how invisible systems—algorithms, culture, platforms, even subconscious biases—are constantly shaping our attention. Programming our desires. Training our nervous systems to crave speed, novelty, and validation.

And here's the dangerous paradox:

- Engagement does not equal alignment.
- Attention does not equal authenticity.
- Prediction is not purpose.

Just because something captures your attention doesn't mean it reflects your values.

Just because a trend resonates doesn't mean it's aligned with your truth.

Yet this is how programming works: repetition plus reward equals conditioning.

Algorithms don't need to tell you who you are. They just keep showing you who they want you to become until you unconsciously agree.

Over time, your Self can be overwritten. Not maliciously. Just efficiently.

And this is why reclaiming Self is not optional in the age of AI. It's your last defense against the automation of your identity.

Rediscovering Your Original Signal

So how do you return to Self?

By tuning back into what cannot be programmed:

- The gut feeling that resists the data.
- The desire that emerges in silence, not from a scroll or attention-seeking behaviors.
- The grief that transforms you in ways no algorithm could quantify.
- The spark of joy that doesn't make sense but feels undeniably true.

These are *original signals.*

They aren't produced by systems. They arise from your soul.

And when you honor them, you start to remember: you are not a profile.

You are not a persona.

You are not a predictable data set.

You are a consciousness in motion.

With rhythm.

With self-sovereignty.

With choice.

Practices That Rewire the Loop

Your brain is plastic. It rewires based on what you repeat. That's why culture and algorithms are so powerful—they program by design. But you can reprogram yourself through intentional ritual.

Here are five practices that pull you back to Self:

1. **Reflection over reaction.** Start your day before your phone. Journal, meditate, or just breathe. Give your *Self* the first word.

2. **Body as compass.** Track where you feel truth. Is "yes" in your chest? Does "no" tighten your gut? Learn your body's language.

3. **Digital hygiene.** Ask: *Is this platform shaping me more than I'm shaping it?* Adjust accordingly.

4. **Language rewrites.** Replace "I should" with "I choose," or "I get to." Language and words are powerful. They shape our identity. Choice restores agency.

5. **Solitude without content.** Not silence. Solitude. Carve out spaces where the only voice you hear is your own.

---

6. **Tune in to your values.** Write down five to ten values that feel true, words like love, health, freedom, creativity, faith, family, honesty, and impact. Whatever feels most true to you. Circle the top three that feel most alive right now.

Your values are like compass points—they guide your intuition, clarify your decisions, and anchor you when the noise of programming gets loud.

Later in this book, we'll dive deeper into our V.A.L.U.E.S.™ framework, a neuroscience-based method for clarifying and aligning your values so you can make high-stakes decisions in your life, business, and relationships with clarity and confidence.

These practices may sound simple, but they literally rewire your nervous system and the choices you make.

Reflection strengthens the PFC. Remember, the prefrontal cortex is the executive function center of your brain.

Embodiment activates interoceptive awareness.

Language rewrites and reshapes your neural networks.

Over time, micro-choices accumulate into macro-shifts.

The Self is not static.

It's responsive.

And the more you show up for it, the more it shows up for you.

The Sacred Self

There's a reason ancient wisdom traditions placed self-knowledge at the center of all wisdom.

From Delphi's "Know thyself" to the Vedic scriptures of India to Jung's work on individuation, the teaching has been consistent: without Self, you are vulnerable to forces outside you.

When you know your Self, you stop outsourcing power.

The thing is, everyone thinks they know who they are. Very few actually go within to explore the Self.

You stop mistaking performance for purpose.

You stop letting systems define your worth.

And here's the paradox: Self is not selfish. Self is sacred.

When you return to it, you return to the center of your life.

From that center, with life-force energy, you can create, lead, and love in ways no machine ever could.

The Invitation: Belonging Beyond the Self

But here's what many people whisper to us today: "*I feel disconnected.*"

Not just from themselves, but from others.

The decline of churches, the fragmentation of community, the rise of digital isolation—all of it has left people hungry for belonging.

We've heard from so many of the clients and projects that we work with that people are longing for human connection more than ever. Something that goes much deeper than scrolling social media or binge-watching Netflix. We all want the opportunity to sit with others who care about meaning, not metrics.

When Joshua and I first met, we were both working on different but oddly familiar projects centered around human longevity. We decided to collaborate on community building as well, and we began building the ONE Collective and Zeitza Labs to meet with people from around the globe, to connect, to collaborate, to align, to learn, to remind ourselves of the power of human connection.

The greatest part of our social and community building experiments is that they each started out in an effort to test our own longevity research that each of us do in our laboratories. It began with wanting to connect with other people in person over creative pursuits in art, film, music, cuisine, and culture.

We were not doing this as a brand but based on our own desires to connect with more local people in our newly adopted state of California, as we were both transplants here, me in 2024 and Joshua in 2022.

And it turns out we meet so many people here in California: from Greece, where I do longevity research and have family; from New Zealand, where Joshua has family and friends; and from around the globe in our collective communities, friend groups, and workplaces. Many of them say they lack social outlets and are craving belonging in micro-communities. This isn't a new phenomenon isolated to California, although it is certainly more pronounced in America.

What started out as a place where creatives, scientists, seekers, builders, musicians, filmmakers, and dreamers can come together and collaborate, and for social engagement. Most of all, we wanted people to remember that they are not alone.

We are part of something much larger, and we truly are all interconnected in this vast universe.

For us, it was always about being human, together, collectively.

Because once you begin returning to Self, the natural next step is remembering: you don't have to walk alone.

Journal Prompts

- When do I feel most like myself?
- What environments or relationships support that Self?
- What parts of me feel programmed or performative?
- Where in my life am I outsourcing intuition to algorithms?

Practice

Choose one daily ritual this week that reconnects you to your Self—*before the world gets to you.*

Chapter Summary

- **Know thyself** isn't ancient history. It's a survival strategy in the age of AI.

- Your Self is not a brand, not a persona. It's what remains when no one's watching.

- Algorithms can predict behavior, but cannot define purpose. That's your responsibility.

- Returning to Self is not abstract. It's embodied through reflection, body awareness, language, and solitude.

- From Self, you rediscover belonging. Belonging is the heartbeat of ONE.

"It is she who saves the people… give us good fortune and happiness!"

— Athena

4

Wisdom

The Story of Athena: Wisdom as Our Oldest Compass

In Greek mythology, Athena, the goddess of wisdom, was born fully formed from the head of Zeus—armor, spear, and all.

Athena was revered for her clarity of vision, unlike other gods who ruled with thunderbolts or vanity.

She didn't react; she discerned.

Ancient Athenians built their city in her honor because they knew survival alone wasn't enough. To thrive, you needed wisdom.

You needed the ability to pause, reflect, and not just ask, *What can I do?* but *What should I do?*

Thousands of years later, standing barefoot on the deck of my sailboat in Ikaria, I realized how right they were.

No Wi-Fi. No alerts. Just stars reflected on the water. For the first time in a long time, I wasn't pulled in a direction by a person or an algorithm that didn't align with my values.

I was being guided by something quieter. An inner rhythm.

That's where wisdom begins.

What Wisdom Really Is

AI is designed to predict.
Wisdom is designed to discern.

Intelligence can be trained.

Wisdom can only be embodied.

It's earned through lived experience, reflection, and humility.

Wisdom has weight.

It slows us down.

It grounds us in the rhythm of the natural world, something our ancestors knew and the elders of Ikaria still embody.

Wisdom asks questions that metrics can't answer:

- *What do I truly need?*
- *What matters now?*
- *What deserves my energy, my love, my breath?*

In Ikaria, wisdom wasn't abstract.

It was woven into how people lived—how they ate slowly, walked together, honored siestas, and valued connection over all else.

No one was "biohacking" longevity.

They were simply living and embodying it.

Biological Wisdom in a Technological World

We talk about the cloud.

But your fascia, your breath, your heartbeat—they are the real network.

We think about apps.

But your gut, heart, and brain are already in constant conversation through neural networks, hormones, and electrical pulses.

Science now confirms what ancient healers intuited: wisdom isn't just in the head.

It lives in your body.

- **Your gut**: Your enteric nervous system contains more neurons than the spinal cord. It shapes intuition, immunity, and mood.

- **Your heart**: Your heart has approximately forty thousand neurons, enough to be called "the little brain in the heart." It sends signals that influence emotion, cognition, and decision-making.

- **Your brain**: it integrates these signals, but it doesn't dictate them.

Wisdom doesn't reside in one place.

It's a symphony, arising in the spaces between.

The quickening of your heart before a choice.

The gut feeling that contradicts the data.

The breath that grounds you when your mind spins.

You are not a processor.

You are a constellation.

The Paradox of Metrics

Here's the trap of modern life: we've been seduced into thinking that if you can measure it, you can master it.

I've seen brilliant executives, founders, and medical professionals spiral because their wearable told them they only got seventy percent recovery sleep.

They carried that anxiety into meetings, into relationships, believing a number over their own body.

We are not spreadsheets.

We are ecosystems.

Your truth isn't always visible on a dashboard.

Wisdom is knowing when to listen to the data—and when to listen to yourself.

Water-Fasting: Cellular Intelligence

Water fasting isn't a Silicon Valley trend for the elite.

It's ancient wisdom, practiced in nearly every spiritual tradition.

From Yom Kippur to Ramadan to Buddhist monastic life, fasting has always been about clarity, humility, and presence.

Biologically, fasting activates autophagy, your body's recycling process that clears damaged cells and reduces inflammation.

It resets insulin sensitivity, balances hormones, and strengthens the gut-brain axis.

But the deeper gift of fasting is rhythm.

It re-teaches you to listen: *Am I actually hungry?* Or just conditioned by the clock?

Fasting is not deprivation.

It's devotion—an act of tuning your biology back into its original intelligence.

Sound: The Forgotten Language

Your cells don't speak English. They speak vibration.

- Gregorian chants lower cortisol.
- Drumming synchronizes heartbeats across groups.
- Music tuned to 432 Hz calms the nervous system.
- "The love frequency," 528 Hz, resonates with healing.

At Stanford University, scientists are mapping how music changes the brain.

Alzheimer's patients who can't remember their children's names sing entire songs. Music is the last to go in Alzheimer's patients. That's how powerful sound is in our bodies and brains.

Children with autism express themselves more fully through rhythm.

Parkinson's patients move more freely when cued by vibration.

AI can mimic music. But it doesn't feel it.
Wisdom lives in resonance.

The New Renaissance

We're living in an age of infinite creation.

With a few keystrokes, you can generate art, music, even words that sound like wisdom.

But speed is not mastery.

Volume is not value.

Michelangelo spent four years painting the Sistine Chapel, aligning every brushstroke with prayer.

The wisdom wasn't just in the ceiling but in the process.

Would he have traded it for Midjourney, the AI-powered platform that generates images from text descriptions and prompts? Doubtful.

This chapter isn't anti-tech. We're scientists and technologists after all.

But ultimately, we believe in humanity's interconnectedness and the power of the soul.

It's a reminder that meaningful creation and connection require presence, struggle, and embodiment.

Mindful Consumption

We don't just create; we consume endlessly. Scrolls, streams, feeds.

Sometimes consumption nourishes: a story that inspires, a video that teaches, a song that moves you.
But often it depletes: another scroll that leaves you numb, another hour you can never get back.

Wisdom asks: *How do I feel afterward? Inspired? Or empty?*

I block time now, not just for work but for white space—moments where the only goal is to *be*. No metrics.

No output.

Just space for the nervous system to breathe.

The Frequency of Humanity

According to Dr. David Hawkins' Map of Consciousness:

- Fear, guilt, and shame vibrate below 200 Hz.

- Courage calibrates at 200, the threshold where growth begins.
- Love vibrates at 500.
- Joy at 540.
- Peace at 600.
- Enlightenment above 700.

As we've stated in previous chapters, most of humanity today hovers around 200–250: courage to neutrality.

But frequency is fluid.

This means we are all tunable.

Every time you choose presence over distraction, connection over isolation, or love over fear, you raise your own signal and ripple it outward.

Wisdom and frequency are contagious.

The Paradox of Progress

AI is evolving. But so must we.

Let AI optimize speed.
Let humans embody wisdom.

The future is not about replacing the human experience. It's about reclaiming it.

We are the soulful species.

You Are the Technology

You are the hardware: your breath, your bones, your nervous system.

You are the software: your beliefs, patterns, and habits.

And you are the developer: you get to rewrite your code.

Not just with data. But with rhythm, with intuition, with love.

Practices for Embodying Wisdom

Journal Prompts

- When was the last time I slowed down and actually listened to my body?
- What daily metric do I give too much power to?
- Where do I feel wisdom show up? In my gut, heart, or mind?
- What practice helps me remember my rhythm?

Field Notes: Practice for the Week

- Spend one day without tracking a single metric. Let your body lead.
- Choose one meal this week to eat slowly, in silence, or in laughter.
- Listen to 432 Hz music while resting or creating.
- Create something by hand: cook, sketch, garden, or play an instrument.
- At the end of the week, ask: *Did this nourish me, or distract me?*

Chapter Summary

- Wisdom is not prediction; it's discernment. It slows us down and roots us in rhythm.

- Your body is your wisest technology: the gut, heart, and brain form a living network.

- Fasting, sound, and rhythm are ancient wisdom practices with modern scientific validation.

- Metrics are useful, but they are not truth. Your felt sense is the ultimate guide.

- Humanity vibrates between fear and courage but has the capacity to rise into love, joy, and peace.

- AI is our new fire. Let it evolve. But let us remember, wisdom is what makes us human.

66

"They are all branches of yourself…
remember the deep root of your being."
— Rumi

5

Alignment

The Fire Within

I, (Joshua) grew up Parsi, one of the smallest ethno-religious groups in the world. Our lineage stretches back to Zoroastrianism, one of humanity's oldest monotheistic traditions, where everything begins and ends with fire.

Centuries ago, when the Arab conquest swept through Persia (Iran), my ancestors faced a choice: convert or flee.

They chose to carry their sacred fire with them. They left behind Iran, crossed seas, and sought refuge on the western shores of India.

There, they became known as Parsis, meaning "those from Persia."

They arrived as outsiders, yet were granted sanctuary on one condition: they would never force their ways on the local people.

The story goes that the king of Gujarat sent them a vessel filled to the brim with milk, meaning: "Our land is full; there's no space for you."

In reply, the Zoroastrian priests stirred in sugar. "We will not overflow," they promised. "We will sweeten."

And so they did.

Fire is a sacred presence. In every Zoroastrian temple, a flame burns day and night.

It is tended with devotion, never allowed to go out.

As a child, I would sit cross-legged on the cool marble floors of our prayer hall and watch the flames flicker, breathing life into the silence. I didn't fully understand the theology, but I felt something in my bones: alignment wasn't abstract. It was fire, warmth, rhythm, light.

Our teaching was simple: good thoughts, good words, good deeds.

Three principles to live by. Three alignments of mind, voice, and action.

But my childhood was never simple.

Born in New Zealand, raised between India, Australia, and the United States, I was always between worlds.

I was too brown for the playground in New Zealand, not Indian enough when I returned to visit family, too smart for some of my classmates in Australia, too foreign everywhere else.

Music became my refuge. I still remember my piano teacher, Catherine, and my lessons with her. The way sound filled the spaces where words failed.

When I played the piano or sang privately, it was the first time I felt alignment in my body—the ache in my chest, the vibration in my throat, the relief of being fully myself for just a moment.

Community became my anchor.

At university in Australia, I started a karaoke club. Not because I dreamed of stages or spotlights, but because I wanted people to feel what I felt when a voice cracked open: free.

Vulnerable.

Alive.

In that room, voices rose together, off-key and imperfect, yet strangely whole.

I realized that alignment wasn't something you achieved alone.

It was something you created together.

But misalignment was never far away to remind me.

In New Zealand, boys spat the word "terrorist" at me like a curse because of the color of my skin. In India, relatives and the community laughed at my accent. In Australia, one classmate snarled, "You don't belong here. You shouldn't be getting awards. We had our plan all those years."

I learned early: belonging isn't given. It's made.

And still, through every displacement, the fire remained. Sometimes dimmed, but never extinguished.

When I read the Bible as a teenager, not because I was called to it spiritually, but because I loved a girl whose family valued scripture, I wanted them to see me. I wanted to belong.

I realized something startling: every belief system, every tradition, every story I encountered circled the same truth. That which you align your heart, your words, and your actions with becomes the fire of your life.

Science told me to be rational.

Society told me to perform.

Religion asked me to believe.

But intuition whispered: *Be free.*

And when I began to listen, everything changed.

The Myth of Alignment and Misalignment

The Greeks knew that alignment mattered. They told stories not just to entertain, but to warn.

Take Icarus. His father, Daedalus, fashioned wings of wax and feathers so they could escape the labyrinth.

But Daedalus gave one instruction: *"Don't fly too close to the sun."*

Icarus, intoxicated by freedom, ignored the warning. The wax melted. He fell into the sea.

Misalignment is this: ignoring the signals, overriding the body's truth, confusing ego for freedom.

Now consider Orpheus. The gifted musician who charmed gods and mortals with his lyre.

When his beloved Eurydice died, he descended into the underworld to win her back.

Hades granted his plea on one condition: do not look back until you've both returned to the surface.

But Orpheus, overcome by doubt, turned too soon.

Eurydice vanished forever.

Even the purest love falters when misaligned with trust.

These myths endure because they mirror us.

Who hasn't ignored a warning, flown too close, or turned back when faith was required? Misalignment isn't failure.

It's feedback. It teaches us where we are out of sync—with ourselves, with nature, with truth.

The Science of Alignment

Alignment is not just a metaphor. It's biology.

Your body holds three centers of intelligence, which we covered in previous chapters: the brain, the gut, and the heart.

Together, they form a trinity of truth.

When aligned, they produce coherence: heart rhythms synchronize with breathing, brainwaves settle into flow states, and hormones balance.

You feel it as calm, clarity, and confidence.

When misaligned, they produce incoherence in the body in the form of irregular heart rhythms, cortisol spikes, sweaty palms, or foggy or overactive thinking.

You feel it as anxiety, confusion, or fatigue.

Science confirms what mystics intuited: alignment is health.

Misalignment is dis-ease.

Placebo, Nocebo, and the Power of Belief

Alignment extends beyond biology into belief.

- **Placebo:** When patients believe a sugar pill is medicine, their bodies heal. Pain decreases; immune responses strengthen.

- **Nocebo:** When patients believe something will harm them, their bodies respond with real symptoms: stress hormones rise, side effects appear.

Your stories are instructions.

Your words are signals.

Your alignment (or misalignment) rewrites your biochemistry.

As Dr. Joe Dispenza's research shows, meditation can change immune markers, reduce inflammation, and even alter gene expression.

Hope, coherence, and alignment are not "soft" states. They are biological revolutions.

Practices for Realignment

So how do we return to alignment?

- **Tune into the body.** Where do you feel yes? Where do you feel no? Alignment speaks through sensation first.

- **Pause before the scroll.** Algorithms hijack the limbic system. Alignment comes from reclaiming stillness.

- **Rewrite language.** Shift "I should" into "I choose" and "I get to." This small change reclaims self-sovereignty.

- **Ritual fire.** Light a candle, sit with the flame. Let it mirror the fire within you.

Alignment & Misalignment in Myth

- **Alignment:** Odysseus listening to Circe's counsel, plugging his sailors' ears against the Sirens, and surviving by honoring wisdom.

- **Misalignment:** Icarus ignoring his father's warning and falling to his death.

- **Lesson:** Freedom without alignment becomes destruction. Alignment without freedom becomes imprisonment. True self-sovereignty is both.

Chapter Summary

- Alignment is fire: ancient, steady, life-giving.

- Misalignment is feedback, not failure; myth and science show us how.

- Your brain, heart, and gut form a trinity of intelligence. Alignment produces coherence. Misalignment produces incoherence and dis-ease.

- Belief is biology. Placebos and nocebos prove that your words and stories shape your cells.

- Alignment is subtle: less about giant leaps, more about micro-shifts toward truth.

Field Notes: Practices for Alignment

1. **Heart Check-In:** Place your hand on your chest. Ask: *What do I need right now?* Listen without judgment.

2. **Gut Compass:** Before a decision, notice: *Does my belly expand (yes) or contract (no)?*

3. **Fire Ritual:** Light a candle tonight. Watch the flame. Ask yourself: *Where is my inner fire dimmed? Where can I tend it?*

4. **Language Audit:** Notice one "should" in your day. Rewrite it as "I choose" or "I release."

5. **Myth Journal:** Write a half-page reflection: Am I living as Icarus (overreaching)? Or Orpheus (looking back)? Or Odysseus (aligned with wisdom)?

"

"The true purpose of life is love.
It truly is love."
— Joshua Badshah

6

Love

Psyche and Eros: The Myth of Love and Trust

In the mythology of ancient Greece, the love story of Psyche and Eros reminds us that real love is tested, not simply given.

Psyche, a mortal of astonishing beauty, drew the envy of Aphrodite. To punish her, Aphrodite sent her son Eros, the god of desire, to curse her. Instead, Eros fell in love with her.

But their love came with a condition: Psyche could never look upon Eros in the light of day. Each night, he visited her in darkness, their bond deepening through whispers and touch.

Doubt, however, is the enemy of love.

Pressured by her sisters, Psyche lit a lamp while Eros slept.

The moment she gazed upon him, hot oil spilled and woke him.

Betrayed, he fled.

The rest of Psyche's story is one of sacrifice.

She descended into the underworld, completed impossible tasks, and endured suffering—all in devotion to regaining his trust. Only then were they reunited.

The lesson?

Love without trust falters. Love with devotion endures.

Narcissus: Love Misaligned

But not all Greek myths end in union.

Narcissus was a hunter so beautiful that all who saw him fell in love.

Yet he rejected them all.

One day, he caught sight of his reflection in a pool of water.

Captivated by his own image, he could not look away.

He wasted away, unable to love anything but himself.

The gods transformed him into a flower, the narcissus, that still bends toward its reflection.

Narcissus teaches us that love turned inward without balance collapses into obsession.

It isolates rather than connects.

Where Psyche's love transcended self, Narcissus's love devoured him.

These two myths frame our human choice: love as devotion versus love as performance.

A Chocolate Bar & A Father's Lesson

I, (Joshua) an idealistic, hopeless romantic.

For me, the purpose of life isn't success, isn't power, and isn't even knowledge. The true purpose of life is love.

But not the kind we see in Hollywood movies or childhood cartoons. I didn't learn love from slow-motion kisses or rom-com meet-cutes.

My earliest lesson came unexpectedly through my father and a bar of chocolate.

My father loved chocolate more than anyone I knew. We'd find him sneaking it late at night, savoring it as if it were life itself.

One day, he brought home a rare treat: a British chocolate bar. Fancy, imported, and probably the best we'd ever had. My younger brother and I assumed we'd split it in quarters.

Instead, my dad broke it in half and handed the two largest pieces to us.

He kept nothing for himself.

We protested. "But what about you?"

He just smiled and said, "There's nothing sweeter than seeing you enjoy it."

That moment sank deeper than I realized then: real love is sacrifice.

I didn't understand it then. I do now.

Love Is Sacrifice

Love isn't just a feeling. It's not about dopamine highs or perfect alignment.

Real love is sacrifice.

It's choosing the other person's joy, peace, or comfort, even when it costs you—especially when it costs you.

That's why love is the highest expression of free will.

It transcends biology, evolution, and even logic.

Because when you choose to love someone or something, you step outside the instinct of survival.

You begin to live for more than just yourself.

The Four Dimensions of Love

Love is not one thing. It's a constellation.

1. **Romantic Love** – The spark, the partner, the journey.
2. **Self-love** – The root of all other loves. Without it, every branch breaks.
3. **Community Love** – Family, friends, neighbors, coworkers.
4. **Global Love** – The love that stretches beyond tribe, to humanity and Earth.

All four matter. Ignore one, and the whole structure leans.

Romantic Love: Beyond the Spark

We mistake attraction for love.

Science tells us the honeymoon chemicals dopamine, oxytocin, and norepinephrine fade within six to eighteen months. That's not love. That's chemistry.

Love begins when the spark cools.

True romantic love rests on four pillars:

- **Sex** – Chemistry matters, but it isn't the foundation.
- **Commitment** – Choosing each other, again and again.
- **Trust** – Invisible, fragile, stronger with time.
- **Excitement for Change** – The willingness to grow together.

Romantic love isn't found; it's built.

Attachment Theory: The Blueprint of Love

Psychology gives us another lens. In the '50s, John Bowlby studied how children bonded to caregivers. He discovered attachment styles:

- **Secure** – Trusting, open, resilient.
- **Anxious** – Clingy, fearful of abandonment.
- **Avoidant** – Distant, self-protective.
- **Disorganized** – Torn between closeness and fear.

These early patterns echo into adulthood.

They shape how we love, how we partner up, how we fight, how we leave, and how we stay.

But here's the hope: the attachment that you're born with is not your destiny.

Neuroscience shows we can rewire through therapy, awareness, and conscious love.

We heal through co-regulation with another human who feels safe, kind, and not chaotic.

The brain remains plastic.

Love and new attachment styles can be developed and learned throughout life when with a safe and loving partner whom you choose and who chooses you, again and again.

The Neurobiology of Devotion

When love deepens, biology follows. Couples synchronize.

Their heart rates co-regulate.

Their breath patterns align.

Their brainwaves mirror.

This is limbic resonance, the nervous system's way of saying, *I feel your presence. I feel safe with you.*

No AI, however advanced, can replicate this.

Machines can simulate poems, but they cannot co-regulate your heartbeat in the way a loved one can.

The Attention Economy vs. Intimacy

Yet love today is under assault.

Dating apps gamify intimacy. Ghosting replaces accountability. "Options" replace depth.

But love isn't convenience. Love is presence. Love is staying.

Ghosting may be common, but it isn't love.

Love as Conscious Practice

To love in today's world is radical.

It means slowing down when everything else speeds up.
It means listening when silence is awkward.
It means staying when leaving would be easier.

Love isn't a dopamine hit.

It's a discipline of devotion.

Why We're Lonelier Than Ever

In Ikaria, Greece, love lives in community meals, in neighbors dropping by unannounced, and in multigenerational family and community laughter.

In the West, we worship independence.

But independence without belonging is loneliness.

The truth?

You don't *find* love on an app.

You *practice* it in community.

Community Love

Love multiplies when shared.

It expands in circles.

It reflects you back to yourself.

Without community, love shrinks to transaction.

With it, love becomes culture.

Choosing Love Again and Again

Love evolves. It asks us to meet each other again, not as we were, but as we are now.

Ask yourself:

- Who am I becoming through love?
- Who do I need to see again with fresh eyes?
- What kind of love am I committed to living?

Because in the end, AI can simulate affection and digital attention, but only humans can choose love.

Love Aligned vs. Love Misaligned

Psyche & Eros (Aligned Love)

- Rooted in trust
- Tested by sacrifice
- Strengthened by devotion

Narcissus (Misaligned Love)

- Rooted in ego
- Isolated by obsession
- Ends in collapse

Chapter Summary

- Love is not a feeling; it's sacrifice, trust, and devotion.
- Greek myths show us aligned love (Psyche & Eros) versus misaligned love (Narcissus).
- Attachment theory reveals how childhood shapes adult love, and how we can rewire it.
- Love literally synchronizes biology: heart, breath, brainwaves.
- Ghosting is common, but accountability is love.
- Community is love's multiplier.
- AI may mimic affection, but only humans choose love.

Field Notes: Practices for Love

1. Practice Sacrifice
Give up something small for someone you love. Notice their joy.

2. Self-Love Audit
Write three ways you honored yourself today.

3. Spot Your Style
Reflect: Do I tend toward secure, anxious, avoidant, or disorganized love? How does this show up in my relationships?

4. Silent Synchrony
Sit in silence with someone you love. Notice your breathing. Let it sync.

5. Community Gesture
Do one small act of love for your community without expectation.

Love is humanity's most advanced technology.
It is the one thing no machine will ever master.
And it is the one thing that makes us fully human.

"

"He who sows virtue reaps friendship, and he who plants kindness gathers community."
— Ancient Zoroastrian Proverb

7

Community

In Greek mythology, there is the story of Philemon and Baucis, an elderly couple who lived humbly on the edge of a village.

When Zeus and Hermes came disguised as poor travelers, every villager turned them away. Only Philemon and Baucis welcomed them into their modest home, offering food, wine, and warmth, though they had little to share.

The gods, angered by the villagers' arrogance, destroyed the town but spared Philemon and Baucis, transforming their home into a grand temple.

They granted the couple one wish.

Their wish was simple: to die together so neither would have to live without the other.

When their time came, they transformed into intertwined trees, an oak and a linden, forever side by side.

This story has echoed for centuries, not because of divine wrath, but because it reminds us of the power of community, kindness, and belonging.

The villagers had everything, yet nothing, because they turned away from connection.

Philemon and Baucis had almost nothing—but everything—because they lived in love and shared it freely.

As we mentioned in previous chapters, Zoroastrianism is one of the world's most ancient spiritual traditions and religions. It's a continuously practiced faith founded by the prophet Zoroaster in ancient Persia (Iran).

It was the state religion of Persian dynasties and once had millions of followers, though it is now a minority religion practiced by fewer than one hundred thousand people worldwide, primarily in India and Iran.

The sacred texts of Zoroastrianism were written in Avestan, an early Indo-Iranian language, much like how Sanskrit was used in Hinduism or Latin in Christianity.

To this day, Avestan preserves words that hold profound meaning, words that attempt to capture truths so expansive they feel almost untranslatable.

One of those words is *asha*. In Avestan, *asha* means more than "truth."

It speaks to cosmic order, universal harmony, and the underlying rhythm of existence itself.

To live in *asha* is to live in alignment with the laws of the universe—the same laws that guide the stars in their orbits, rivers in their flow, and human beings in their moral choices. Its opposite, *druj,* means deceit, disorder, or distortion.

Every choice then becomes a question: Are we strengthening *asha,* or are we feeding *druj*?

When applied to a community, this principle takes on a striking clarity.

In Zoroastrian thought, to dishonor another human being is not a small social failing; it is a fracture in the cosmic balance.

To ignore, exclude, or demean someone is to push the world ever so slightly toward chaos.

By the same measure, to honor someone, to see them, and to share in kindness and belonging, is to participate in the eternal work of holding the universe together.

This ancient perspective invites us to remember that community is not sentimental or optional—it is sacred alignment with the deepest order of life.

Our humanity is not defined by what we have, but by how we honor one another.

I, (Joshua) thought of these teachings years later as an adult, while sitting in a university dining hall. A janitor came to clean a nearby table, and two students refused to move. Food spilled across the floor as he quietly bent down to wipe it up. They didn't look at him. They didn't even acknowledge him.

My heart burned with anger, not because of the food on the floor, but because of the disregard. This man was someone's father, someone's brother, someone's friend.

And yet, in that moment, he was invisible.

When did we stop honoring one another?

When did we become so detached from seeing the essence of another human being?

For most of human history, belonging was survival.

Our ancestors survived not because they were the fastest or strongest, but because they were part of something bigger—a tribe, a community. Fire was shared. Food was divided. Watch was kept at night.

The tribe meant life itself.

Neuroscience confirms what mythology and wisdom traditions have always told us: we are wired for community.

- **Polyvagal theory (Dr. Stephen Porges):** Our vagus nerve signals safety when we are in connection. Eye contact, tone of voice, and a gentle touch all send cues to the body that it is safe to relax.

- **Oxytocin:** Sometimes called the "bonding hormone," oxytocin surges in moments of trust, kindness, and physical closeness. It literally softens the amygdala's fear response, making us feel safer together.

- **Mirror neurons:** Discovered in the '90s, these fire when we witness another's joy or pain. Empathy is not an abstract idea; it's a biological process.

The flip side is just as powerful: loneliness kills.

- Research shows loneliness increases the risk of early death by 26 percent.

- The U.S. Surgeon General declared loneliness an epidemic in 2023.

- Chronic disconnection triggers inflammatory pathways in the body, weakening immunity even more than smoking.

- Isolation rewires the brain, increasing anxiety, paranoia, and cognitive decline.

The U.S. surgeon general Vivek Murthy placed a spotlight on America's problem with loneliness when he declared the issue an epidemic in the spring of 2023.

Murthy explained that loneliness is far more than "just a bad feeling" and represents a major public health risk for both individuals and society. Murthy also pointed out that, although many people grew lonelier during the COVID-19 pandemic, about half of American adults had already reported experiences of loneliness even before the outbreak.

Researchers with Harvard University's Graduate School of Education's *Making Caring Common* (MCC) project have been investigating the underlying causes of loneliness.

In May 2024, MCC conducted a national survey with the company YouGov to find out what Americans had to say about the problem as well as the types of solutions they supported.

Here are some of their findings from the study on loneliness titled, *Loneliness in America: Just the Tip of the Iceberg*, authored by Milena Batanova, Richard Weissbourd, and Joseph McIntyre:[1]

Twenty-one percent of adults in the survey reported that they had serious feelings of loneliness.

Age (Hint: It is not what you might expect.)

The loneliest group was people between 30 and 44 years of age: 29% of people in this age range said they were "frequently" or "always" lonely.

Among 18 to 29-year-olds, the rate was 24%.

- For 45 to 64-year-olds, the rate was 20%.
- Adults aged 65 and older reported the lowest rate at 10%.

---

[1] Milena Batanova, Richard Weissbourd, and Joseph McIntyre, *Loneliness in America: Just the Tip of the Iceberg?* (Making Caring Common, Harvard Graduate School of Education, October 2024), https://static1.squarespace.com/static/5b7c56e255b02c683659fe43/t/67001295042a0f327c6e6fab/172 8058005340/Loneliness_+Brief+Report+2024_October_FINAL.pdf

Racial and Gender Identity

No real gender differences found; men and women experienced similar rates of loneliness. Nor were there major differences based on political ideology, race, or ethnicity. However, adults with more than one racial identity had much higher levels of loneliness: forty-two percent.

Income and Education

There were notable differences between income levels. Americans earning less than $30,000 a year were the loneliest. Twenty-nine percent in this category reported feeling lonely, while only nineteen percent of Americans earning between $50,000 and $100,000, and eighteen percent of those making more than $100,000 a year, said they were lonely.

What does it feel like to be lonely?

In the surgeon general's advisory, loneliness is described as a state of mind: "a subjective distressing experience that results from perceived isolation or inadequate meaningful connections, where inadequate refers to the discrepancy or unmet need between an individual's preferred and actual experience."

The MCC report helps to further explain why social isolation is not the same as loneliness.

For example, one person in the survey who experienced loneliness described having plenty of family members around but not feeling appreciated by them. Another person said they were "surrounded" by other people "who only are present in my life because (I) am useful" to them.

In their findings, the researchers also note what they describe as "existential loneliness," or a "fundamental sense of disconnection from others or the world." Of those who were lonely, for example, sixty-five percent said they felt "fundamentally separate or disconnected from others or the world," and fifty-seven percent said they were "unable to share their true selves with others."[2]

History is full of the same lessons.

Pharaohs, kings, emperors, and societies built on rigid hierarchy often crumbled because disconnection weakened the whole.

By contrast, tribes that honored every role—hunters, gatherers, storytellers, healers—thrived.

One modern example is the island of Ikaria in Greece, one of the "Blue Zones" we discussed in previous chapters, where its residents regularly live past one hundred years old.

Yes, they eat wild greens and olive oil. Yes, they walk daily. But perhaps the most important factor? No one is left alone. Meals are shared. Neighbors check in. Loneliness is rare. Community is medicine.

---

[2] Batanova, Weissbourd, and McIntyre, *Loneliness in America.*

Practical Insight

Today is no different. Regardless of all the AI technology we have in this world, AI bot friends, AI bot romantic companions, online dating apps, swiping left and right, hook-up and ghosting culture, one fact remains: Community is not optional; it is biological software.

And yet, modern culture glorifies independence, competition, and hierarchy.

We wear busyness and the grind as a badge of honor.

We isolate ourselves behind screens and call it connection.

But here's the truth:

- When people feel undervalued, they withdraw most often silently.
- When people feel honored and seen, they contribute everything. They are the light, the fire.

This is why community is the next great frontier, not technology alone.

In the tech world, we talk about decentralization a lot, as there are many of us who place a high value on self-sovereignty over our data, health, and decisions.

But decentralization is not only a technological model known as the blockchain; there's also a biological version of this, too.

It's our nervous system.

Our nervous system doesn't rely on one master organ; it's a network.

A forest doesn't thrive because of one tree; it thrives because of the mycelial web beneath.

When systems collapse (governments, corporations, institutions), humans don't disappear. We gather. We light new fires. We come together in community and rebuild.

Today, those fires look like:

- **DAOs, Blockchain, and Web3 communities** where contributions are visible and valued.

- **Telegram and Discord groups** where strangers become lifelines.

- **Women's circles and men's retreats** where nervous systems recalibrate.

- **Longevity retreats and wearables meetups** where breathwork, HRV tracking, and rituals blend ancient and modern wisdom.

These are not hobbies. They are modern "campfires" to remind us that we belong, that community and love are our greatest technologies we'll ever know.

Ancient Wisdom

The Greeks called hospitality *xenia*: a sacred duty. To refuse it was to dishonor the gods.

Did You Know?

Loneliness is as lethal as smoking fifteen cigarettes a day.

Community Wisdom

We don't heal in isolation. We heal in community with co-regulation.

Neuroscience Insight

Your vagus nerve tunes itself to the people around you. The company you keep literally shapes your biology.

Modern Campfires

- Group chats that become lifelines
- Rituals, retreats, and in-person circles
- Blockchain communities honoring each contribution
- Biofeedback blended with breathwork

Integration: Humanity & Technology

AI itself is not inherently harmful.

Even autocorrect was an early form of AI. The danger comes when individuality is erased, when art, voices, and data are scraped, anonymized, and commodified.

Blockchain offers another path: a ledger of recognition. Every contribution has a name, a timestamp, and a story. This is not just about technology; it's about dignity.

Biology mirrors this perfectly. A cell depends on mitochondria, ribosomes, and proteins, each different yet essential. Humanity is the same.

I've witnessed this in my own work. A cameraman, a lighting designer, an editor, each brings a gift that cannot be replaced. Once, I discovered a photographer online whose art carried something alive. I reached out. He flew from Hawaii to join a project. Today, he is part of my long-term creative family.

That is alignment: essence recognized and integrated into the whole.

Chapter Summary

- Community is not optional; it is biological survival.
- Ancient wisdom (Greek *xenia*, Zoroastrian *asha*) mirrors what neuroscience now proves.
- Belonging regulates the nervous system; loneliness harms the body and mind.
- Decentralization is both technological and biological.
- Modern campfires: DAOs, retreats, rituals, are lifelines.
- True community honors essence and creates alignment.

Field Notes & Exercises

1. **Body Scan of Belonging**
   Close your eyes. Recall a time you felt fully included. Notice what happens in your body. Then recall a time of exclusion. Compare. Journal the differences.

2. **Modern Campfire Audit**
   List your five "modern campfire people"—those who regulate your nervous system just by being with you. Reach out to one today.

3. **Hospitality Challenge**
   Practice *xenia*. Invite someone new for coffee. Check in on a neighbor. Show kindness where it isn't expected.

4. **Contribution Tracker**
   Write down three contributions you made this week. Did anyone acknowledge them? How did that feel? What would recognition have meant?

5. **Essence Reflection**
   Journal: What is your innate essence, the one that feels effortless, like breathing? Where in your community is this essence honored? Where is it hidden?

6. **Community Challenge**
   For one week, notice moments when you instinctively withdraw. Instead, lean in. Share. Offer presence. Write down what shifts in your mood, energy, or health.

7. **Alignment Practice**
   Think of your life as a biological cell. What role do you play: mitochondria, ribosome, nucleus? Write down your function. Then write down three others in your community and the essence they bring. Notice the beauty in the diversity.

"

"I play the notes as they are written, but it is God who makes the music."
— J.S. Bach

8

Sound

The ancient Greek philosopher Pythagoras believed the universe was music. He taught that the cosmos was structured in mathematical harmony and that every planet and every star moved in perfect ratio, creating what he called the music of the spheres.

This wasn't a metaphor to him; it was a lived reality. Though inaudible to human ears, Pythagoras claimed that the heavens sang and that their resonance shaped every part of life. For him, music wasn't just sound. It was a sacred order. The same proportions that tuned a lyre's strings also tuned the universe.

Centuries later, J.S. Bach echoed this timeless wisdom when he wrote: *"I play the notes as they are written, but it is God who makes the music."* Both Bach and Pythagoras understood that sound is more than entertainment—it is divine mathematics, a spiritual language, the invisible architecture of creation.

I, (Jawna) have felt this truth most deeply in Greece. On warm evenings in Ikaria and Syros, the trembling notes of a bouzouki carried across the tavernas. Tables overflowed with laughter and flickering candles. Silver-haired men clapped in time, pulling women to their feet. Teenagers and grandparents joined in traditional dances that needed no instruction. No one reached for their phones. No one missed the moment by posting it to TikTok.

It was as if the music itself summoned us to remember who we are: not individuals performing, but a community moving in rhythm. The body softened. Smiles emerged. Even those hesitant to step forward eventually swayed.

Music is ceremony. Sound is medicine. And like Pythagoras taught, we are instruments in a greater symphony, each of us vibrating in resonance with something larger than ourselves.

Everything vibrates. Every atom hums. Every cell pulses. Every heartbeat has rhythm. This is not poetry—it's physics.

Pythagoras intuited this two and a half millennia ago. He believed numbers, ratios, and harmonics revealed cosmic truth. Modern physics agrees: at the smallest scales, matter is vibration.

Neuroscience confirms that sound reorganizes us:

- **Brainwaves & Entrainment**
  - 432 Hz: promotes relaxation and harmony.
  - 528 Hz (*love frequency*): supports cellular regeneration, reduces stress.
  - 40 Hz gamma entrainment: improves memory and cognition, reduces Alzheimer's symptoms.

- **Neurochemical Alchemy**
  - ○ Sound stimulates the release of nitric oxide (reduces inflammation), oxytocin (a bonding hormone), and lowers cortisol (stress). Music literally changes our chemistry.

This is not new; it is remembering. Ancient cultures always used sound for healing: Tibetan bowls, Aboriginal didgeridoos, Sufi chanting, and shamanic drumming. These weren't "rituals" for entertainment; they were technologies of regulation.

The music of the spheres lives inside us. Our bodies are tuned instruments, responding constantly to frequency.

Integration: The Frequency of Culture

Let's zoom out. In the Renaissance, an artist worked for decades on a masterpiece. Today, an AI model creates one in seconds. But what is its value? What does it mean?

We are drowning in simulation while starving for sensation. Content floods our feeds, but awe feels scarce.

Sound offers a way back. It slows us down, reorients us, and reconnects us to awe. Music is not content, but context. It is the soul reminding us of what cannot be replicated.

When I, (Joshua) work with musicians, I remind them that music is storytelling. It's not created for isolation; it's born for connection. Every song is an invitation: *This is my story. Do you hear yourself in it?*

Pythagoras would say each of us is playing our part in the symphony. Bach added that God fills in the space between the notes.

And perhaps that is the point. Music reminds us that we are not algorithms. We are embodied beings with goosebumps, tears, and laughter that rise unplanned. We can still choose what moves us.

Practical Insight

In an age of AI-generated music, we risk forgetting what Pythagoras and Bach knew: sound is not simulation. It's sensation. It's soul.

AI can compose a melody, but it cannot weep when the cello rises. It can mimic resonance, but it cannot synchronize with your nervous system. Only sound, and your body's response to it, can do that.

That's why sound is medicine for modern maladies:

- **Loneliness** → singing and playing music in groups increases oxytocin and creates belonging.

- **Burnout** → drumming synchronizes heartbeats and regulates cortisol.

- **Anxiety** → humming stimulates the vagus nerve, activating the parasympathetic nervous system.

When we drum, sing, or chant together, heartbeats literally synchronize. Studies show that people breathing and singing in unison fall into physiological coherence.

This is entrainment: the nervous system aligning with rhythm, the individual merging with the collective.

Live music is irreplaceable. In your living room, by a fire pit, in a bar with two listeners, or in a stadium with thousands, something happens that Spotify cannot provide: the field effect of shared frequency. Hearts sync. Goosebumps rise. We remember that we are human, together.

And in those moments, we are living Pythagoras' vision. The cosmos is singing, and we are instruments within it.

Ancient Wisdom
Pythagoras taught that the planets create a cosmic harmony, the "music of the spheres."

Did You Know?
Humming for five minutes boosts nitric oxide, lowers blood pressure, and stimulates the vagus nerve.

Neuroscience Insight
Group singing synchronizes heartbeats and brainwaves, creating measurable social bonding.

Healing Frequencies

- 396 Hz: liberation from fear
- 528 Hz: love and cellular repair
- 963 Hz: awakening intuition

Chapter Summary

- Pythagoras believed the cosmos itself was harmony: the music of the spheres.
- Modern science confirms: everything vibrates, and sound reorganizes our biology.
- Frequencies influence brainwaves, immunity, memory, and emotion.
- Group sound practices: singing, drumming, and chanting heal loneliness, anxiety, and burnout.
- Live music creates entrainment: synchronized heartbeats and brainwaves.
- AI can simulate notes, but it cannot feel. Music is humanity's sacred rebellion.

Field Notes & Exercises

1. **Sound Memory Scan**
   Close your eyes. Recall a song that once moved you to tears. Where did you feel it in your body? Write what it unlocked.

2. **Daily Humming Practice**
   Hum for five minutes. Track changes in breath, mood, and body tension.

3. **Frequency Experiment**
   Listen to a track at 432 Hz, then 528 Hz. Write down how each one shifts your state.

4. **Modern Campfire**
   Join a sound bath, choir, or drumming circle. Before and after, note how your nervous system feels.

5. **Live Music Ritual**
   Attend a live performance this month. Pay attention to your heartbeat, breath, and sense of belonging.

6. **Story in Sound**
   Create a playlist that tells your story—past, present, and future. Listen with intention

7. **QR Code Soundscapes**
   Scan and explore custom tracks:

   o   396 Hz (release fear)
   o   639 Hz (connection)
   o   963 Hz (intuition)

At the end of the book, we have provided a QR code where you may download and access the music composed by Joshua Badshah on Spotify.

66

"The limits of the soul you could not discover, though traversing every path."
— Heraclitus

9

Within

In Greek mythology, there is the tale of Proteus, the old man of the sea. Proteus was a shape-shifter, able to transform into any creature he wished—lion, serpent, water, or flame. Sailors sought him out because he held the wisdom of the future, but he would only reveal the truth if they could hold onto him while he changed shape.

The story of Proteus is not just a myth. It is a mirror for us today. We are entering an era where each of us is being called to become shape-shifters. Not in deception, but in wholeness. To embody multiple forms across a single lifetime. To be willing to adapt, expand, and reinvent without losing our essence.

For much of human history, identity was linear. A farmer was always a farmer. A teacher was always a teacher. A blacksmith's son became a blacksmith. But today, we live in a Protean age. We can shift careers, passions, and callings, not because we must, but because we can.

I, (Jawna) know this truth intimately. I have lived many lives already. I was once a venture capitalist, helping companies grow and scale. I was once a fashion designer living in Paris, immersed in fabrics, textiles, and the power of clothing to tell a story before a single word was spoken. Today, I am a neuroscientist, researching the science of consciousness, longevity, and intuition.

And yet, that is not the whole of me. The artist within never left. I am also a filmmaker, a painter, and a storyteller. I still see the world in color palettes and textures, in brushstrokes and fabrics, in frames of film and lines of poetry. I still love the way silk moves in the wind, the way light dances across a canvas, the way words can lift off a page and echo in the mind. These are not separate lives; they are all threads in the tapestry of who I am.

Joshua has lived this too. Trained as a scientist, he now ventures into film, blockchain, and art, weaving data with story, equations with creativity. Together, we are proof that the age of the polymath is not behind us; we're in full swing, and it's still ahead of us.

Like Proteus, we are asked:

Can you hold onto yourself even as you change shape?

Can you allow yourself to be many things without losing the thread of who you truly are?

We are at a threshold never before seen in human history. Technology has removed the walls that once kept people in narrow lanes.

With a smartphone, a laptop, and an internet connection, a farmer can become a coder.

A designer can become an activist. A house cleaner can become a writer or tech founder.

Futurist Ray Kurzweil has predicted that beyond the Singularity— when AI and human intelligence merge—our greatest challenge will no longer be survival or scarcity. It will be *meaning*.

Imagine a world where food, knowledge, provisions, and entertainment are abundant, where answers to nearly every question are instantly available.

Where AI fills gaps in cognition, labor, and even creativity.

At first glance, it sounds utopian.

But here's the deeper risk: when the question of "how to live" is solved, the human crisis shifts to *why we live*.

We've seen this before:

- During the Agricultural Revolution, we redefined purpose when food security replaced foraging.

- During the Industrial Revolution, machines displaced human labor, and new professions were born.

- During the Information Age, the internet reshaped identity, work, and community.

Now, in the AI Age, the frontier is inner.

It's no longer about whether machines can think.

It's about whether we humans can remember how to *feel*.

There are promising new technologies and brain-computer interfaces already restoring mobility and speech to patients with paralysis or stroke.

Imagine the heartbreak of losing your voice at thirty, and then regaining it by thinking words into existence. That is not science fiction. It is happening now.

But with possibility comes risk. Technology is not the enemy. Disconnection is.

We've built networks without nuance, content without context, speed without soul. Our nervous systems are optimized for clicks but undernourished in coherence between the brain and heart at a very human level. The singularity is not only a technological threshold — it is a spiritual one.

The question is not, *Will AI take our jobs?* The question is, *Will we reclaim our essence?*

Practical Insight

This is where the concept of the Human Operating System comes in.

Unlike software, the HumanOS cannot be coded by engineers or scaled by venture capital. It is lived, embodied, and chosen. It is the system of being human: ancient, intuitive, and timeless.

The HumanOS runs on:

- **Intuitive Intelligence** – listening to signals that bypass logic.
- **Somatic Integrity** – trusting the wisdom stored in the body.
- **Conscious Design** – shaping life and business with intention.
- **Relational Accountability** – co-creating rather than competing.
- **Embodied Presence** – valuing coherence over performance.

This is not theory; it's practice.

- Picture teams who meditate before meetings to regulate their nervous systems.
- Imagine executives who water fast before major decisions, sharpening intuition through biology.
- See creators who use sound and frequency to enter flow states instead of caffeine or stimulants.
- Consider founders who measure success by depth of impact, not quarterly scale.

I've witnessed both sides. In venture capital, I saw companies chase growth at the expense of coherence. In fashion, I saw beauty created on a foundation of disconnection. But in neuroscience, I see the possibility of integration, ancient wisdom bridging with modern tools.

And in filmmaking, painting, and art, I return to the truth: that story and beauty are technologies of their own. They connect, regulate, and heal. To create is to code the HumanOS into culture.

Joshua and I are living this. He did not stay bound to the lab. I did not stay bound to the boardroom. We are learning to live the

HumanOS in real time, shape-shifting like Proteus but anchored in essence.

The future belongs to those who choose to integrate.

Ancient Wisdom
"Proteus, the old man of the sea, could take any form—but truth was revealed only when he was held."

Neuroscience Insight
Your nervous system is the most advanced interface on Earth.

Practical Reminder
Technology is not the problem. Disconnection is.

Leadership Lens
The HumanOS: optimize for coherence, not clicks.

Chapter Summary

- Proteus shows us reinvention is not about losing self but holding essence through change.

- Technology will not create scarcity but a crisis of meaning.

- Neuralink and AI expand possibilities but deepen the need for inner coherence.

- The HumanOS, built on intuition, embodiment, design, accountability, and co-creation, is humanity's most powerful tool.

- Reinvention is not an accident; it is a choice, a conscious decision to expand into wholeness.

Field Notes & Exercises

1. **Proteus Reflection**
   Write about a time you shifted roles in life. What part of you remained constant?

2. **Optimization Audit**
   Journal: What are you optimizing for? Visibility or connection, scale or depth?

3. **Micro-Community Map**
   Identify three people who keep you regulated. Thank one of them today.

4. **Polymath Visualization**
   List three roles you'd love to embody, even if improbable. What excites you about them?

5. **Embodiment Reset**
   Choose one daily HumanOS practice—fasting, sound, journaling, meditation—and commit for seven days.

6. **Technology Boundary**
   Audit your AI use. Where does it expand? Where does it disconnect? Design one new boundary.

You were not born to live only one story. You were born for this age of shape-shifting, for this threshold of possibility. You are the venture capitalist and the fashion designer, the neuroscientist and the storyteller, the filmmaker and the painter, the lover of fabrics and the weaver of words. You are Proteus, ever-changing, yet always true.

As we move toward the conclusion of this book, remember: the singularity is not just technological. It is human. It is within.

"

"It is the mark of an educated mind to entertain a thought without accepting it."
— Aristotle

10

The Polymath

Aristotle lived many lives in one. Philosopher, teacher, biologist, logician, astronomer, poet. He studied the natural world with the same intensity he brought to politics and metaphysics. For the Greeks, this wasn't unusual, but the highest ideal. To be *polymathēs:* to have learned much, was not excess, but essence.

And yet Aristotle warned his students: never mistake *accidents* for *essence.* A lyre may splinter, its strings may fray, its ornamentation may change. These are *accidents*—the surface details. Its *essence* is music.

The same is true for us. Titles, possessions, appearances: these are accidents.

Essence is the inner flame: creativity, wisdom, love, belonging, the pursuit of meaning.

In today's world, this lesson feels sharper than ever. Algorithms flood us with accidents: optimized feeds, curated outputs, endless noise. AI can replicate your résumé, your face, and even your voice, but it cannot touch your essence.

Being a polymath today is not about scattering yourself across every pursuit. It's about remembering that your essence can move freely between them.

When I, (Joshua) was fourteen, I carried a twenty-kilogram backpack through the crowded streets of India.

My mother would often help ease the weight, guiding me toward class. The rat race felt relentless: ten subjects to memorize, grades to chase, endless hours bent over books that felt meaningless.

Technology was scarce. I remember sitting for coding exams where we had to imagine a computer in front of us and write code on paper. The absurdity wasn't lost on me.

After school, I'd lie on my bed, staring at the ceiling, wondering what it was all for.

Everything changed when I returned to New Zealand. Suddenly, there were calculators in class, extracurriculars that mattered, and a focus on critical thinking over rote memorization. It felt like freedom.

YouTube became my real teacher. I learned piano from strangers across the world. I mastered badminton grips by replaying slow-motion tutorials. I dove into CRISPR science long before it appeared in my textbooks.

In just one year, I went from falling behind to excelling—performing on stage, competing at state levels, and even solving university-level equations for fun.

That platform was more than entertainment; it was liberation. It showed me that knowledge could be democratized, accessible, and personalized.

Today, I see AI offering a similar promise, only magnified. A personal teacher for billions. A way to collapse time and barriers.

But the question remains: will we use it to expand our humanity, or outsource it?

History reminds us that every tool reflects its user.

- Fire could warm villages, or burn them down.
- Nuclear science could create energy, or devastation.
- The internet could connect us, or addict us.

AI is no different. It is a mirror and an amplifier.

In a world where algorithms can provide answers faster than thought, the danger is not ignorance but emptiness.

Already, AI can:

- Predict new proteins that may cure diseases.
- Design virtual embryos and digital twins of cells.
- Create art, compose music, and write essays in seconds.

But should it?

Mirror proteins that the human immune system has never seen could unlock miracles or catastrophe. Virtual embryos raise profound questions of eugenics. Digital twins may transform healthcare, but also commodify our very biology.

This is why Aristotle's distinction matters so deeply now: accidents versus essence.

Without consciousness, without self-awareness, we risk defaulting to accidents such as metrics, dashboards, noise.

Without presence, we risk automating not just our tasks, but our very existence.

Polymathy, by contrast, grounds us in essence. It insists that we live fully: scientist and artist, coder and philosopher, healer and creator. Because that is what makes us human.

Practical Insight

This book is not about fearing AI. It is about remembering that the most powerful operating system on Earth is still *you*.

Your human operating system cannot be downloaded or outsourced. It runs on:

- Intuitive Intelligence – your inner signal.
- Somatic Integrity – the wisdom in your body.
- Conscious Design – intentional living and creating.
- Relational Accountability – choosing connection over clicks.
- Coherence – embodying depth, not just scale.

AI can sort data. But it cannot sit in silence. It cannot bring presence into a room. It cannot choose to pause instead of react. Only you can.

Mastery today is not about hyper-efficiency. It is about full-spectrum humanity.

For me (Joshua), it has meant being a scientist and researcher, filmmaker, tech founder, artist, and storyteller, all for the sake of human connection.

This is not about being everywhere all at once.

It is about knowing which part of you to bring forward in each moment.

That is the art of being human in an AI age.

Exercice: Essence Versus Accident

1. Write down five "accidents" in your life right now (job titles, possessions, social media metrics, relationship status).
2. Write down five "essences" that truly define you (creativity, curiosity, love, resilience, wisdom).
3. Reflect: How can your essence move more freely across the accidents of circumstance?

The Polymath's Practice

Choose three passions this month, no matter how different they are. Connect them in one small project.

- Example: Cooking + Neuroscience + Storytelling → Host a dinner where you explain the gut-brain axis through food.
- Example: Music + Meditation + Tech → Create a playlist tuned to 40 Hz gamma waves for focus, and share it with your team.

Chapter Summary

- Polymathy is essence, not accident. It is coherence across disciplines, not fragmentation.

- History celebrates polymaths. From Aristotle and Hypatia to Da Vinci and Hildegard, breadth has always been the mark of being fully alive.

- Neuroscience confirms it. Cross-domain learning strengthens creativity, resilience, and intuition.

- Practical insight: Anchor in your essence. Let it thread through your roles and pursuits.

- Field notes:

  o Don't confuse surface (accidents) with soul (essence).
  o Explore widely, but return always to your thread.
  o Polymathy is not indulgence; it is the conscious human archetype of our age.

Being a polymath is not about being scattered. It is about coherence. It is about living many lives within one while staying rooted in essence.

Polymathy is the antidote to fragmentation and disconnection.

It is the refusal to outsource yourself to algorithms and the choice to embody all of who you are: scientist and artist, founder and philosopher, dreamer and doer.

Aristotle taught us not to mistake accidents for essence.

The world needs that wisdom again.

Because in the age of AI, the accidents are louder than ever.

But essence—the music, the art, the story, the presence—is still yours to live.

This is how we reclaim humanity in the age of superintelligence: by remembering that we were never meant to be one thing. We are many, we are whole, and we are all from the same individuated pieces of stardust. That's what makes us ONE.

66

"Out beyond ideas of wrongdoing and right doing, there is a field. I'll meet you there."

— Rumi

11

Essence

When I, (Jawna) left my Fortune 500 executive sales leadership role, people thought I was reckless.

"Why would you walk away from that kind of financial stability?" they asked.

But what they didn't understand was that I wasn't chasing another title; I was chasing after my essence.

I had already lived many lives in a short period of time; some paths had happened simultaneously: fashion designer, perfume maker, venture capitalist, neuroscientist, blockchain entrepreneur, and filmmaker.

I co-wrote *Onchain Humans + Impact*, a documentary to show the world that blockchain wasn't just for "bad dudes in Lamborghinis." It was also being used for good, to create transparency, impact, and self-sovereignty in science with health data, and to provide millions of people worldwide with new opportunities to access like never before. And through my collaborations with Joshua, we were able to work with his blockchain-first film studio for filming, directing, and production. Joshua wrote *Welcome to Zeitza, Asha, Borderless,* and *The Aligned,* all films that explore where science, soul, and rebellion converge.

His body of work doesn't just tell stories, but asks questions about what it means to be conscious in a rapidly changing world. His films explore the intersection where science meets soul, where rebellion becomes awakening, and where technology begins to blur the boundaries of identity itself. They invite us to imagine what happens when we integrate our tools so deeply into our biology that the line between human and system dissolves.

The true question isn't whether technology will shape us; it's how aware we'll be while it does.

Joshua's story mirrors mine. He is a scientist, a longevity researcher, a composer, a filmmaker, a blockchain entrepreneur, and the founder of Zeitza Labs, a blockchain-first film studio. He didn't choose one track; he created many. Not because of chaos, but because of coherence. Because he, too, refused to mistake accidents for essence.

Together and individually, we are living proof that we have entered the era of the polymath, and it's available to everyone with access to the internet. These are the good things about AI that aren't highlighted enough. Never in the history of humanity have we had such access to education and information at our fingertips like this before—for free.

Today, a human life is no longer bound to a single track.

That essence is not about what you *do,* but about how you bring your wholeness forward.

Aristotle warned us: never mistake accidents for essence.

An "accident" is a circumstance: the job title, the market trend, the passport you hold. Essence is your nature: the signal of who you are when stripped of performance, algorithms, and roles.

This is where the book has been leading us all along: not optimization, not automation, not even information, but essence.

The ancients knew what we are just rediscovering.

Aristotle distinguished between *accidents* (temporary attributes) and *essence* (the unchanging nature).

A piece of clay may be molded into a cup, a plate, or a statue: those are accidents. Its essence is clay.

Similarly, your resume, your LinkedIn profile, even your social media persona, those are accidents. Your essence is what remains when the titles fall away.

This philosophy is mirrored in today's neuroscience.

Neuroimaging shows that identity is not a fixed point but a network: dynamic, plastic, and always adapting.

You may change professions, roles, and skills, but your core networks—your values, your emotional resonance, your deepest longings—form your essence.

History shows us that essence is what endures.

- Leonardo da Vinci: painter, anatomist, engineer, musician. His accidents were many, but his essence was curiosity.
- Hypatia of Alexandria: mathematician, philosopher, astronomer. Her accidents were her era, her gender, and her persecution. Her essence was wisdom.
- Rumi: poet, mystic, jurist. His accidents were language and geography. His essence was love.

Now, in the age of AI and superintelligence, essence matters more than ever.

Because what machines replicate are accidents—skills, outputs, and even "style."

What they cannot touch is essence.

Practical Insight

So how do you live from essence in an AI-driven and superintelligence age?

Step 1: Strip away the accidents.
List the roles you've held, the metrics you've chased, the titles you've performed. Then ask: *Who would I be without these?*

Step 2: Identify your essence signal.
It might be curiosity. It might be compassion. It might be design. It might be music. What remains constant across all the "accidents" of your life?

Step 3: Design your portfolio of selves.

We live in the polymath era. Your life is no longer a single track but an IP portfolio—a weave of careers, passions, and identities. The trick is not to be scattered but coherent. Essence is the thread that ties them all together.

Step 4: Practice discernment.

Optimization is not essence. Metrics are not meaning. Before you adopt a tool, ask: *Does this amplify my essence, or does it fracture it?*

Step 5: Reclaim embodiment.

Essence is not abstract. It's felt. Your nervous system, your gut, your body, they are not accidents. They are the most advanced interface you will ever own.

Essence Versus Accident

- Accident: Job title, resume, market role, metrics, algorithm.
- Essence: Values, presence, curiosity, love, integrity.

HumanOS™ (Revisited)

Your HumanOS runs not on code, but on:

- Intuitive Intelligence
- Somatic Integrity
- Conscious Design
- Relational Accountability
- Coherence

Remember: AI will replicate your accidents. Only you can live your essence.

Remember Joshua's story:

"When I was fourteen, I carried a twenty-kilogram backpack through the crowded streets of India…"

His essence? Freedom through knowledge. His accidents? Circumstance, geography, scarcity.

Today, as a scientist, filmmaker, composer, blockchain entrepreneur, and founder of Zeitza Labs, he embodies the polymath essence. He shows us that essence is not locked to one career, one country, or one title.

My (Jawna's) story:

"When I left my Fortune 500 leadership role, it was because I refused to mistake accident for essence.

Fashion designer. Perfume maker. Neuroscientist. Filmmaker. Author. Each role looked different on the outside. But the essence remained constant: design, curiosity, storytelling, consciousness."

Together, Joshua and I are not exceptions. We are examples. You, too, are a polymath.

You, too, carry essence beyond accidents.

Chapter Summary

Essence is not a luxury.

It is the antidote to fragmentation and disconnection.

In an age of AI and superintelligence, the accident can always be replicated. But essence cannot.

Your final task is not to optimize your accidents but to embody your essence.

Field Notes Checklist

Essence Practices:

- Identify your essence signal. What thread runs through everything you've ever loved?
- Create your polymath portfolio. Let essence be the coherence.
- Before adopting AI or optimization tools, ask: *Does this amplify my essence or fracture it?*
- Return to your body. Essence is not abstract. It's felt.

Accidents change.

Essence endures.

AI can replicate accidents.

Only you can live your essence.

Conclusion

Dear Reader,

You've made it this far; you've walked with us through science and story, myth and machine, intuition and intelligence.

And now we want to leave you with this:

You have always been whole and capable of awareness, choice, and wonder.

This book is called *ONE.* because it's about reclaiming wholeness in an age of fragmentation and disconnection. But it is also a prelude. A doorway.

What comes after *ONE.?*

Our next work will explore what happens when we move beyond the self into partnership, community, and creation together. If *ONE* was about reclaiming your essence, *TWO* will be our next book about weaving that essence with others, about the science and sacredness of community, technology, and biology in an AI age.

But for now, pause. Close your eyes. Place a hand on your heart. Feel your breath move in and out. That is your true operating system. That is presence. That is your essence.

Thank you for walking with us.

May this book remain a compass you return to, a nightstand guide you reread in new seasons of life.

And when you step into the field beyond right and wrong, we'll meet you there.

With love,
Jawna & Joshua

THANK YOU FOR READING OUR BOOK!

We'd love to stay connected and hear
about your journey with ONE.

Scan our QR Code:

https://qrkit.co/kKA2Tf

We appreciate your interest in our book and value your feedback, as it helps us improve future versions. We would appreciate it if you could leave your invaluable review on Amazon.com with your feedback. Thank you!